TECHNOWORLD
Going Digital

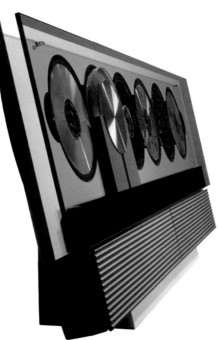

 RAINTREE
STECK-VAUGHN
PUBLISHERS

A Harcourt Company

Austin   New York
www.raintreesteckvaughn.com

Published by Raintree Steck-Vaughn Publishers,
an imprint of Steck-Vaughn Company

Library of Congress Cataloging-in-Publication Data

Graham, Ian 1953-
    Going digital/Ian Graham.
      p. cm--(Technoworld)
    Includes bibliographical references and index.
    ISBN 0-7398-3254-9
    1. Digital communications--Juvenile literature. 2. Digital electronics--Juvenile
literature.[1. Digital communications. 2. Digital electronics.] I. Title. II. Series.

    TK5102.4 .G75 2001
    621.382--dc21
                             00-062824

Printed in Italy. Bound in the United States.

1 2 3 4 5 6 7 8 9 0 LB 04 03 02 01 00

All screen shots in this book appear courtesy of the producers of the
websites they depict and remain the copyright of those producers.

All trademarks are acknowledged.
Apple, the Apple logo, Macintosh, QuickTime and any other Apple
products referenced herein are trademarks of Apple Computer, Inc.,
registered in the US and other countries. Eudora is a registered
trademark of the University of Illinois Board of Trustees, licensed to
Qualcomm Incorporated. Microsoft, Microsoft Internet Explorer, MSN,
and any other Microsoft products referenced herein are registered
trademarks of Microsoft Corporation in the US and other countries.
Netscape Communications Corporation has not authorized, sponsored,
endorsed or approved this publication and is not responsible for its
content. Netscape and the Netscape Communications corporate logos
are trademarks of Netscape Communications Corporation.

CONTENTS

THE DIGITAL WORLD `4`

HOW DIGITAL DATA WORKS `6`

THE FIRST DIGITAL SYSTEMS `8`

GOING DIGITAL `10`

THE DIGITAL LIFE `22`

THE DIGITAL FUTURE `28`

GLOSSARY `30`

FURTHER READING `31`

INDEX `32`

Everything is going digital. From music, games, and photography to telephones, video, and television, a digital revolution is sweeping across the world.

Digital machines aren't new. Computers have been digital since the 1940s, but now other machines are going digital, too.

Diaries and address books have gone digital. Electronic organizers store names, addresses, telephone numbers, and other information.

Digital Does It Better

Doing things digitally is popular because digital is better. Digital sounds and pictures suffer from less of the crackles, hisses, and interference that spoil telephone calls, recorded music, radio, and television.

Digital Designs

The digital revolution now makes it possible to design something in one country, e-mail the design to another country so that it can be tested, and then e-mailed to yet another country, where the object is actually made, perhaps by computer-controlled robots.

▲1. A mountain bicycle is designed in England.

▲ 2. The design is sent to the United States by e-mail for a series of tests using computers.

▶ 3. The design is then e-mailed to Korea where the bicycle is built by robots. The manufacturers and designers keep in touch by e-mail and telephone.

Digital means using electronic digits or numbers. Information that digital machines handle, whether it's sound, pictures, or words, is changed into numbers, processed, and then changed back into sounds, pictures, or words again.

Wave Machines

Before digital machines came along, most electronic devices were analog machines. Analog machines change sounds and pictures into electric currents or radio waves. The microphone in a telephone is an analog device. It changes your voice into a vibrating electric current that grows larger or vibrates differently as you speak.

Sending Digital Data

A microphone changes sounds into a vibrating electric wave. The wave is an electrical copy of the sounds.

The size of the wave is measured and changed into a series of "slices."

Each slice of the electric wave is given a number. Taller slices are given bigger numbers.

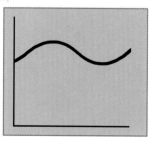

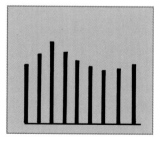

Why Bother Going Digital?

A television set uses the radio waves it receives to create pictures. If the radio waves are damaged, perhaps by a thunderstorm, the picture suffers, too. A digital television set receives radio waves that represent a code. The TV set receives the code and uses it to create the pictures. As long as the code gets through to the TV set, no matter how damaged it is, the TV set can still "read" the code and use it to create perfect pictures.

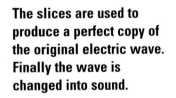

The numbers are changed into a code of zeros and ones called binary code. The code is transmitted as a series of pulses.

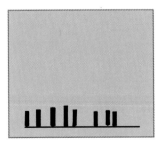

When the pulses are received, they are changed back into the numbers that give the height of the slices.

The slices are used to produce a perfect copy of the original electric wave. Finally the wave is changed into sound.

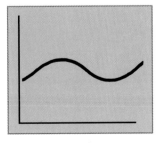

The first digital computers were built in the United States, Great Britain, and Germany in the 1940s. They each weighed several tons and filled entire rooms.

Univac was the first electronic computer built by a private company. It was made in 1951 in the United States.

Digital Communications

Telephone calls were connected digitally for the first time in Britain and in the United States in the 1970s. The first digital mobile phone system was set up in the United States in 1992.

Fiber Optics

Telephones used to be connected by metal wires carrying electric currents. The wires are now being replaced by glass cables called fiber optics. Calls travel along them as beams of light. In addition to telephone calls, fiber optics can also carry computer data and images.

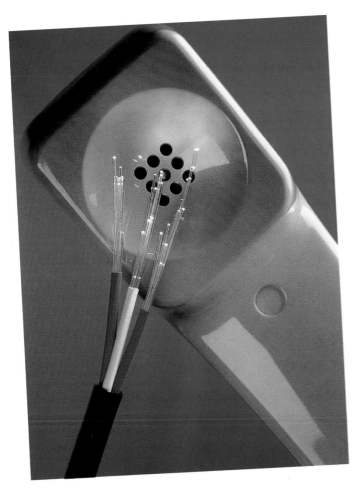

Telephone calls travel as light beams down thin glass fibers. The beams stay inside the strands even if they bend around corners.

Inside a Mouse

When you move a computer's mouse, a ball underneath it rolls and turns wheels inside it. The computer works out how the mouse is moving by counting how many times the wheels turn and then makes a pointer on its screen move in the same way.

Most of the electronic equipment we use today is going digital.

Computers

A computer is a super-fast calculator. It does millions of calculations every second. A set of instructions called a program tells the computer what to do with the answers. It can change them into sounds, pictures, or words.

Computer Networks

A number of computers connected together is called a network. Computers can be connected together by telephone using modems to change each computer's digital information into electric currents that can be sent down a telephone line.

Pictures can be sent by e-mail. A picture is placed on a scanner. This changes it into digital information that can be fed into a computer. Then it is sent to someone else's computer by e-mail and can be printed on paper.

Surfing the Web

The biggest computer network is the Internet. Using it, you can send messages called e-mails to computers all over the world. And on the World Wide Web, you can find all sorts of information on electronic pages called web pages. A group of web pages is called a website.

▶ Each web page is linked to other web pages. Mouse-clicking on a link takes you straight to another page.

MUSIC BY NUMBERS

Recorded music sounds better than ever, thanks to digital recording, digital instruments, and digital sound systems.

▶ **A CD can hold up to an hour of music. It is recorded as millions of pits "burned" into the disc. When light bounces off them, they make the reflection flicker, and this is changed into sound.**

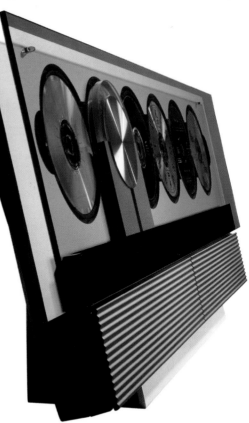

Digital Discs and Tape

The most popular digital disc is the compact disc, or CD. The digital code on the disc is "read" by firing a laser beam of light at it and turning the flickering reflection into sound. There is also a type of digital tape called Digital Audio Tape (DAT).

A Digital Audio Tape (DAT) machine makes digital recordings on a tape cassette.

On-line Music

You can download digital music from an Internet website and play it on your computer or a portable player called an MP3 player.

Digital Radio

In the future, radio programs will be sent out to everyone as a stream of digital code. Radio sets with digital decoders will turn the digits into sound. Radio programs will no longer suffer from crackles or hisses, but will be as clear as a CD.

An MP3 player doesn't have a tape or disc inside. It loads music into its memory directly from the Internet.

A Digital Versatile Disc (DVD) player can play discs like CDs that may contain computer data, music, or full-length motion pictures.

IN A RECORDING STUDIO

Recording studios make digital copies of music. Different musical instruments and voices are recorded separately. Then the recordings are combined and refined to make the final mix.

1. Musicians are recorded separately as they play their instruments and sing in the soundproof studio.

2. A machine called a sampler takes sounds recorded from the real world, such as waves on a shore or a clap of thunder, and can change the sound into digital code. The sound can then be played on a keyboard at different pitches.

3. A machine called a mixer adds all the separate recordings together to make the final recording.

Hundreds of knobs and switches on the mixing desk control the way each instrument and voice sounds.

4. The final, or master, recording is copied onto a disc or digital tape.

DIGITAL TELEVISION

A television picture is made from hundreds of lines of light (525 in the United States and Japan, 625 in Europe). In the future, digital TV sets will show better pictures made from double the number of lines as today. This is called High Definition Television, or HDTV.

Digital television is bringing us dozens of new television channels.

Decoders

A digital television receives radio waves that translate into a stream of binary numbers. A computer inside the TV changes them into a picture. We can watch digital television today on an "ordinary" analog television set by using a decoder. It changes the digital information received by aerial or cable so that the television set can use it.

Interactive TV

Digital television sets connected to a telephone line can let you do your shopping and banking from home, play games, and send e-mails, by keying information into a handset. When information can be sent as well as being received, this is called interactive television.

▶ **Interactive TV lets viewers send information as well as receive programs.**

▼ **Games can be played interactively on TV.**

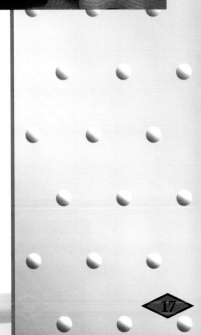

TELEPHONES

Telephone calls are clearer than ever because of digital technology. A telephone changes a caller's voice into a wave-like electric current, which travels to a telephone exchange. There, it is changed into a stream of pulses, a digital code, and sent on its way by cable and radio. Another exchange changes it back into an electric current and sends it on to another telephone.

▲ Telephone calls used to be connected manually by switchboard operators. Callers told them the number they wanted, and the operator connected their telephone line to it.

◀ In a modern telephone exchange, telephone calls are connected in a fraction of a second by computers.

ISDN

A different type of telephone line, an Integrated Services Digital Network (ISDN) line, is digital all the way from one end to the other. It can carry all sorts of digital code, including telephone calls, computer data, and video pictures. ISDN is three times faster than a normal modem connection, and is often used by businesses.

Mobile Phones

Mobile phones send and receive calls as digital radio signals using a nearby radio aerial. If the phone moves too far from the aerial, the call is automatically switched to another aerial.

Mobile phones are connected to the rest of the telephone network by radio.

TAKING DIGITAL PICTURES

The latest digital cameras take photographs without film and store them in a chip or on a magnetic disc. When you take a picture with a digital camera, the light falls on a computer chip. It turns the light into digital code, which is stored in another chip or a magnetic disc.

Once a digital picture is taken, it can be loaded onto a computer and then printed out at any size, as many times as you like.

Film Code

"Old-fashioned" photography with film is also going digital. The pattern of black and silver squares on the side of a roll of film is a digital code that a camera can "read." The silver squares touch contacts inside the camera and tell it how much light the film needs when a photograph is taken. An Advanced Photo System camera (APS) records information on film that a developing machine can use to produce the best photographs.

▼ **Digital cameras look like other cameras, but they have computer chips inside them instead of film.**

Pixels

A digital picture is made from hundreds of lines of colored spots called pixels. Smaller pixels make better pictures, because the tiny spots "merge" together and make pictures that look more like photographs.

THE DIGITAL LIFE

Digital technology is now a part of our everyday life. Everything—from computer games to home appliances to security systems—is going digital.

Better Safe than Sorry

Passwords, codes, alarms—information stored in computers can be stolen just like information printed on paper. Digital locks are used to keep people from breaking into computers. To open the lock, the computer has to be given a secret code, called a password. Digital locks are used to protect all sorts of things, from personal computers and house security systems to cars and safes.

Some computers will not work unless they are given a secret code called a password.

Digital Keys

Some cars are locked and unlocked using a digital code. Pressing a button on the car's key sends a radio signal to the car. The signal contains a number. If the car receives the correct number, it unlocks the doors.

This car key contains a tiny radio transmitter. Pressing a button on the key switches the transmitter on.

If the radio message from the key contains the correct code, the car operates its electronic door locks.

Private Calls

The first mobile telephones used radio waves that anyone could pick up and listen to. Today's digital mobile phones turn a caller's voice into a digital code before transmitting it, so that no one can listen into a call.

DIGITAL FUN

Computer games are streams of digital code that tell a computer to create pictures of cars, planes, or people on the screen and move them around according to how you press keys or move a joystick. Computers today can process code faster than ever, so the pictures are more realistic.

▲ Today, computer games are increasingly realistic. Cars drive through a landscape that exists only inside the computer.

On-line Games

You can download games (copy them into your computer) from an Internet website and play them. Or you can play on-line games against other people who are connected to the Internet at the same time as you.

Virtual Reality

Virtual-reality games are designed to look like a real world. Players wear a helmet with screens in front of their eyes so that they can see only the virtual world made by the computer.

▶ ▼ **Every movement of a player's body causes some kind of reaction in the game.**

EVERYDAY DIGITAL

Lots of machines in your home are already digital. Washing machines, video recorders, microwave ovens, and many other household appliances are controlled by digital circuits just like computers.

On the Cards

We often use plastic cards to buy things. One way of doing this is to "swipe" the card through a machine that reads digital information stored in the card and charges the cost of things bought to the shopper's account.

Money Machines

A bank cash machine works by reading information in a cash card's magnetic strip. The owner also keys in a code, a personal identification number (PIN). The machine sends the information digitally to the card owner's bank. If it is correct, the machine gives out cash.

THE DIGITAL FUTURE

A computer program can make a computer produce all sorts of sounds, including the sound of human speech. Now, there are computer programs that can do something much harder—make a computer understand what someone says to it. The computer "learns" to recognize a person's own speech patterns.

A disabled person can control a computer by speaking into a microphone instead of having to use a keyboard.

TECHNOFACT

Optical Computers

Future computers may be connected together by beams of light instead of wires. These optical computers will work at the speed of light.

Good Morning, Plane!

The new Eurofighter plane has a computer that its pilot can control by talking to it instead of having to press switches.

People Patterns

Fingerprints, patterns on the iris (the colored part of the eye), and patterns of blood vessels at the back of the eye are different for every person. These patterns can be turned into digital code inside a computer. In the future, you might unlock a door by simply looking at a computer or placing a finger on a pad so that the computer can make sure it's really you by checking your eyes or fingerprints.

As even more machines start using the same digital "language" as computers, it will be easier to make machines do exactly what we need them to do and even communicate with them. The future is digital.

The Eurofighter's flight computer listens to what its pilot says and carries out his or her instructions.

GLOSSARY

aerial A metal rod or dish used to send or receive radio waves.

analog A system that uses ever-changing electric currents or radio waves to carry information.

audio Another word for sound.

binary Using numbers made only of a series of zeros and ones.

bit A binary digit, a number that can only be zero or one.

circuit A pathway made from electrical devices connected together so that electric currents can travel between them.

code Letters, numbers, or signs that represent information.

data Information, especially information used by computers.

decoder A box that changes information from one form to another.

digit A number.

digital Using digits to carry information.

download Copy information from a website into another computer.

e-mail Short for "electronic mail," messages sent from one computer to another, using the Internet.

electric current A flow of electricity.

fiber optic A cable made from strands of glass that carries information, such as a telephone call, on a beam of light.

Internet A computer network that stretches around the world.

modem A box that helps to link a computer to a telephone line so that it can communicate with other computers.

network A group of computers linked together so that they can share programs or information.

on-line Connected to the Internet.

password A secret word needed to use some computers or computer programs.

PIN Personal Identity Number, a secret number needed to use a digital machine such as a bank cash machine.

pixel Short for "picture element," one of the thousands or millions of spots that a digital picture is made from.

program A list of instructions that tells a computer what to do.

radio wave A wave of energy that can carry information through space at the speed of light.

World Wide Web A huge library of pages of information stored in computers all over the world, linked together by phone.

FURTHER READING

Hoare, Stephen. *20th Century Inventions: Digital Revolution.* Wayland, 1998
Hughes, Lisa. *The Internet.* Hodder Children's Books, 1998
Wright, David. *Inventors & Inventions: Computers.* Benchmark Books, 1996

WEBSITES

www.digitalcentury.com/encyclo/update/comp_hd.html
www.ipl.org/youth/
www.robertniles.com/data
www.ala.org/parentspage/greatsites/science.html

Picture Acknowledgments:
The publishers would like to thank the following for permission to reproduce their pictures:
8 Archive Photos/Image Bank, 9 Phillip Hayson/Science Photo Library, 18 top NMPFT/Hulton-Getty/Science & Society Picture Library, 18 bottom Andy Snow/Science Photo Library, 25 top Science Photo Library, 25 bottom Peter Menzel/Science Photo Library, 27 Steve Horrell/Science Photo Library, 28 Peter Menzel/Science Photo Library.

INDEX

analog systems 6, 8, 16

binary numbers 9

cameras 20, 21
cash machines 27
compact disc (CD) 12, 13, 15
computers 4, 5, 8, 9, 10, 11, 13, 18, 19, 22, 24, 25, 26, 27, 28, 29
credit cards 27

decoders 16
Digital Audio Tape (DAT) 12, 15
Digital Versatile Disc (DVD) 13

e-mail 5, 10, 11, 17
Eurofighter 29

fiber optics 9

games 4, 17, 22, 24, 25

High Definition Television (HDTV) 16
home appliances 26

Integrated Service Digital Network (ISDN) 19
interactive television 17
Internet 11, 13, 24

mixer 15
mobile phones 8, 19, 23
modem 10, 19
mouse 10
MP3 player 13
music 4, 12, 13, 14, 15

passwords 22
photography 4, 20, 21
pixels 21

radio 4, 13, 18, 19, 23
radio waves 6, 7, 16, 23

sampler 14
scanner 10
security systems 22, 23
smart cards 27
synthesizer 14

telephone 4, 6, 8, 9, 10, 17, 18, 19
television 4, 7, 16, 17

video 4, 19
virtual reality 25

World Wide Web 11, 24